If you are

planning to join the military in any capacity that could put you in a combat zone;

or have recently enlisted in the military;

or you are a combat veteran or anyone else interested in U.S. military history—

then this report is for you.

But you don't have to fall into any of these categories to find this report of great interest.

Here are three examples:

This book has a lot of practical tips for survivalists, military enthusiasts and people who like to camp.

From a historical standpoint, anyone wanting to know more about our war on terrorism in Iraq through the eyes of an Infantry soldier will enjoy this read.

Civilian businessmen and women could use this report as the American version of Sun Tau's *The Art of War*, published in 1913, and still studied by corporate Japan.

Author's Note

Rules of an intelligence agent:

    Admit nothing
    Deny everything.
    Make counter accusations.
    Demand proof.

The good news

If you survive your tour, by God's Grace, you'll have stories to tell your grandkids.

The bad news…

…well, don't think about that. Pessimists do not belong on the battlefield, especially when the insurgents can smell fear like a fart at a party. I'm going to give it to you straight from a grunt's point of view.

My Tour of Duties were primarily Asia and the Middle East. However, the experience I've gained is also valuable to anyone stationed or contracted anywhere in the military ranks. For as many years as I've been an Infantryman (trusted, busted and trusted again) the combat missions were nothing like I'd ever expected.

Training is important, but what gets you through is adaptability. Up was down over there, left was right, and circles were squares. There were battles going on all around us, some of the most powerful were waged from within.

This manual is a little unorthodox and there is very little structure to it, which is fitting to the subject matter. Foreign soil makes hardly any sense to the average Westerner, and the people we fight have absolutely no clue about our culture except that they envy it. So much so that some are set to destroy all freedom with terrorist acts. A lot of violence occurred because of a newly-liberated Iraq, and if there is even one paragraph of this manual that will contribute to your survival on the battlefield (or garrison) then my efforts were worth it.

Please take any knowledge here and from your own training, even though it may <u>still</u> not be enough. Think. Wherever you are. Think, but not too much. Your reactions are crucial. Two things a Special Forces Operative told me before I left:

"If something doesn't look right, it probably isn't."
The second, "If something isn't right, shoot it."
His buddy added, "It's better to be tried by twelve than carried by six."

Some of the following chapters will contain a story about a crucial moment in my tour. Then we'll go over what I did (or didn't do) so that you don't make those same mistakes. In this book are many short, brisk chapters that may imprint your subconscious more effectively than the less and thicker chapters

would. As I just mentioned, you want to retain as much of this as possible when chaos strikes. God was very good to me. More than a few stupid choices I made during my military career should have left me kicking the ol' oxygen habit. Instead of re-inventing the wheel, read this report from cover to cover, and any others on the subject. If you've already been deployed and the bullets are flying too soon for you to finish every page, read Chaplain's Chat near the end of this report.
With that being said, Forward March.

Sebastian DiGiovanni
September 2012

**Survive Your Tour
In Peace & War**

Sebastian DiGiovanni

Published by UCS PRESS
UCS PRESS is an imprint of MarJim Books
P.O. Box 12797
Prescott. Arizona 86304-2797

web site: http://www.marjimbooks.com

Copyright 2012 by Sebastian DiGiovanni

Cover design by Marty Dobkins

All rights reserved.

**This report may not be reproduced in any form
without written permission from the publisher.**

ISBN 978-0-943247-58-8

This unofficial field and training guide is for informational purposes only.

**For God, Hydie, our Kiddos, Family,
Jim, Marty, Larry and Country.**

Acknowledgments

Thanks:

To a few of those mean, arrogant, selfish, trivial bosses we had to put up with out there in training and the slop of real world missions. You taught us a few things, but didn't have to be such jerks when it wasn't necessary. News flash, idiots ... We now have a volunteer military, so lay off the old war movies to try to find your sense of character. Anyway, no hard feelings, but please get help. Also would like to thank all those brainwashed socialists and lifelong college students who patronized me on my return. Yet another reason to write this report.

Many Thanks:

To all the men and women who serve or have served, combat or not. If it wasn't for you brave folks, this report (not to mention our freedom) wouldn't be possible.

Special Thanks:

Sergeant Major of the Universe Jesus Christ, my wonderful partner Hydie, our Kiddos and Family, UCS PRESS, 1st Sgt B (the Pope of the Army), Ram, E, Munchumba, 368th Intelligence, 503rd Air Assault, 101st Airborne and the 40th Infantry Division ... BALLS DEEP!

Table of Contents

Chapter 1: Before the Siege . . . 6
Chapter 2: Godspeed . . . 8
Chapter 3: Fight or Flight . . . 9
Chapter 4: Kings for a Year . . . 11
Chapter 5: Riding the Trigger . . . 12
Chapter 6: Press . . . 13
Chapter 7: Pouges and Rouges . . . 14
Chapter 8: Traitors . . . 16
Chapter 9: Pissing Contests . . . 16
Chapter 10: Emergency Hypnosis . . . 17
Chapter 11: The Art of War . . . 19
Chapter 12: Bootleggers 'R Us . . . 19
Chapter 13: Understanding Terrorism . . . 21
Chapter 14: Support System . . . 22
Chapter 15: Everything but the Sink . . . 22
Chapter 16: Going Home . . . 23
Chapter 17: Over-Vigilance . . . 27
Chapter 18: Chaplin's Corner . . . 27
Chapter 19: Reuniting with Family . . . 28
Chapter 20: My words of farewell . . . 31
Meet the Author . . . 31
Table #4: Bill's Companion . . . 32

Chapter 1: Before the Siege

What can I say about close and long range combat to prepare you?

Nothing.

Crapsolutely nothing.

That's because all the training in the world won't prepare you for that one battle in which your mind will blank out all training and you either stand your ground and fight, or retreat (as we will discuss more thoroughly in the Fight or Flight chapter). One battle isn't enough to turn you pro, yet it will ultimately pop your cherry as a Combat Infantryman. So let's use the training my Company endured as a sample for you to know what to expect. We had it the worst, due to the fact that we were a National Guard unit attached to a Regular Army unit. We were labeled dirt-bags even before those guys met us—this opinion we soon disproved to them. They didn't factor that many of us had regular army experience. I was one of many who were actively deployed to other regions of the world and operating under assignments still deemed classified before entering the Guard. A few of us had already served in combat.

To even be considered for war, you will have to provide and fill out a lot of paperwork. This could take weeks to process, if not months. Things like a valid driver's license for one, as well as birth certificate, social security card and an extensive inquiry through a government or contracted website, listing everything you've done since birth.

Make sure your references are legitimate, as the Department of Defense does check EVERY application with a lice comb.

Next is medical. If you want to deploy, keep your answers short. You have to be healthy and physically fit to pass such an exam.

You'll go through a number of medical briefings before even laying a hand on any sort of weapon, except to clean `em. Answer truthfully. Yet if you complain about any back or knee injuries, guess what? You're not going. Talk about the pain in your stomach that you thought might have been an ulcer, you'll still be enlisted, but not on the front lines. Only way you're deploying is if you can prove to these doctoral folks that you won't be the weak link in time of war. Otherwise, you will be deemed a liability, and they'll stamp your papers that way as a 'medical hold.' You'll have to remain stateside until they find a slightly useful task for you that won't involve violating your profile. But whatever they choose will still suck.

Now please remember that whatever group you get sent with, be it military or civilian, you may get stuck with a bunch that think they're real cowboys. It happened to me. I couldn't stand my squad leader from the very beginning. The guy thought he was John freakin' Wayne. This dude was a garbage man by trade, with an Honor Guard tab on his uniform. He took pride in his monthly responsibilities to the Guard, and everything had to be perfect to make up for his wasted civilian alter ego. Let's call him Sergeant X. His next in command was an unpleasant man who used Jedi philosophy and screaming to get his point across. Let's call him Sergeant Y.

Since I had the most time and grade among the lower enlisted of our squad, Sgt X offered me a team leader position, which I promptly turned down. I didn't want the extra headaches of being in charge of three other troops, especially if I wasn't offered hard stripes to go with it. The National Guard is notorious for tasking soldiers out with extra duties, and I knew this well after spending about two years with a Mortar detachment just a few cities away.

I hardly knew these two new Sergeants, but the company they were attached to was gearing up for war, so I volunteered to join them. But an easy-going demeanor ended up getting on Sergeant X's bad side right from the start. Which brings me to my first lesson learned:

Before deploying make sure to be enthusiastic in front of your superiors.

There will be a LOT of training to endure before you get to any war. Give 100% through every road-march, while cleaning your weapon, whatever. First impressions are crucial and you can always take it down a notch when you land in country. If you get some hard-ass leader that offers you more work, just take it. Just don't let them dump on you. If you truly don't feel comfortable in the squad you are assigned, find one you like and negotiate a trade with someone from that squad. Then each of you can go to the Platoon Sergeants and plead your case. Be careful not to mud-sling against anyone that outranks you this early in the game (I'll teach you about that in a later chapter.) If it's an even swap, it won't be a big deal. Seasoned Platoon Sergeants know that some personalities just happen to clash and they will usually try and accommodate the platoon as a whole before everyone gets settled.

You'll really only get one (or *maybe* two) of these chances during your entire time with these people, so research the squad you are requesting before opening your mouth. Make sure this new squad has room for you, and you will be more content there for the rest of your training and tour. Be selective with any changes you make in the company. This goes for Regular Army, Guard/Reserve or civilian contractor. The last thing you want is to be labeled a whiney bitch before you've even been deployed. Find a buddy to watch your back. "Battle Buddy" is the Army's term for it. Find someone in your squad, preferably with the same rank, and look out for each other. Not only are these bonds invaluable in war, but will probably become like a brotherhood during your tour. These guys will sometimes get on your nerves but would also do anything to protect you, right or wrong. This only works if you would do the same for them, as it should.

Be tactful with your family's questions.

They will be terrified from what they hear on the news, anyway. Try not to paint as bad a picture for your family when you are told what to expect in your mission briefings. Just tell your family you are prepared and well trained for what lies ahead, and always downplay whatever the Army says about the war you may be facing.

Make sure you don't bring too much civilian crap with you. Everything you need for your stay will be issued to you. I'd suggest a few sentimental knick-knacks and pictures in a small backpack or suitcase along with your hygiene kit, one pair of *civvies* and civilian shoes, gum, cash, Gameboy, and this unofficial military guide.

There will be so many things to buy in country. People you don't even know from the states are going to send all kinds of crap. Things like food, books, razors, candy, Q-tips, soap, hand deodorizer, toilet paper, and everything else.

Getting back to being prepared, you will develop more skill than you ever did in Basic Training or Boot Camp. We were sent especially to Fort Lewis to train with the Special Forces. It was training in the worst conditions imaginable. The old Infantry saying "If it ain't raining, we ain't training" rang true. Two months of wearing MILES gear over our basic load and shooting blanks at former Iraqi insurgents who were now working as training aides. They would laugh at our mistakes as if this whole thing were a big joke. Maybe it was a big joke for them. Bastards. Our entire company decided to shave our heads into mo-hawks as an act of defiance from the unforgiving conditions of training. On one occasion we refused the mess hall's inedible slop as we chanted "fuck your chow" upon exiting the facility. After our miserable training was over, California Gov. Swartzenegger flew up to Washington to give us a warm speech. He called us the "real terminators" and shook everybody's hand as we boarded the plane to a new land of terror…

Chapter 2: Godspeed

The planes we hop are civilian ones, rented out by the Army, and everything onboard looks the same as civilian coach—right down to the stewardess. The only difference is all the passengers are soldiers and things can get a little rowdy. If you've ever flown on a commercial airliner as a civilian you will enjoy the contrast of the flight attendants' announcement here:

"All soldiers please make sure your rifles are pointed muzzle down, and grenade pins and clips are securely fastened, as well as your seat belts." I had to chuckle when thinking back to the time I almost missed my flight as a civilian due to a small pocket knife in my bag. As a soldier they let me board this 747 with a machine-gun, ammo and some fragmentation grenades. Cool!

When the overhead lights went out as the plane prepared for take-off, a few soldiers pull out colored mini lights, making designs on the ceiling as if we were in Caesar's Palace. The stewardess has already been hit on 17 times and the plane hasn't even left the ground. Make sure to drink if you are allowed. Don't drink enough to draw attention to yourself or get in trouble, but enough to at least catch a buzz. You won't see the "pain-go-bye-bye juice" for the rest of your tour in country if you play by Army regulations. Save Christmas, when you are allowed two beers—Budweiser or Bud Lite—and you have to sign for them as if you were signing for your weapon. The government has strict drinking laws concerning alcohol in the Middle East, especially any combat zones, on who gets drunk. If you are carrying a weapon and have name-tapes it probably won't be you.

After a sixteen-hour flight, we were on the move again. Everyone that gets stationed in Iraq and all the other arteries of the slop will have to bed down for at least a night in Kuwait.

Take your boots off and change your socks. Soldiers that are too lazy to take their boots off this early in the game usually get a nasty case of trench foot during their tour of duty. Don't be one of those nasty troops. The showers will be warm, so take full advantage of them. You probably will not have a shower that long for a while. No matter what military dentists tell you, once every twenty-four hours is not an acceptable amount of time to brush your teeth unless you want to have yuck mouth when you get back home to the states. Once in the morning and once at night is a good rule, that's why mom taught you early in life. Start your good hygiene habits now, because in the field you will need even more discipline to take care of your body.

The water is bottled for a reason.

Every year of this war the Army spends millions on bottled water. There is a reason for it, because the Army is usually cheap unless there is necessity. Any last swig of H2O you enjoyed before the desert is something you can look forward to in six months to two years from now. Don't take a chance with the disgusting sludge they call water in this unfamiliar continent. There are many parasites in the easterner's water that you don't want to take a chance with. Don't even gargle with it. Get in the habit of keeping your mouth closed when taking a shower.

The first time a little Middle Eastern water got in my mouth I had uncontrollable diarrhea for about two weeks. There are still claw marks on the outhouse walls from that crap episode of mine. It felt like my bowls were a microwave oven containing a boiling hot caldron of black-eyed beans before I was able to let that steaming hot mess out of my system. Multiply Mexico's "Montezuma's Revenge" by five hundred million if the tap water touches your insides. I qualified expert with two light machine guns and a 203 Grenade Launcher during training stateside, so I thought my aim was pretty good. But let me tell you I didn't even make it into the cheap plastic bowl of the makeshift latrine with during a confrontation with the liquid magma pouring through me like a Yosemite hot spring during the rainy season.

Tension will be high among the chain of command before reaching your destination.

Just do what they say and avoid arguments. The truth is, Kuwait is full of pouges (sissies) that walk around talking tough, as if Kuwait were under siege. They will share stories about how the post gets hit with artillery and other attacks, which is total crap. Kuwait is a staging area completely secured by US forces on the border. Nevertheless, high ranking officials who have never stepped beyond the wall and into a combat zone will scare the unseasoned parts of your chain of command into believing the enemy sleeps right outside the front gates. Sure, I'll give them the fact that insurgents get closer to the wall every day, but it's nothing to be alarmed about until you reach the border of Kuwait and the other countries that lie in wait. From there, your entire convoy will pass an American checkpoint with "Godspeed" spray painted in big florescent letters on a cement slab for all gunners pointed at three o'clock to see. Beyond those gates, the real chaos of war begins.

Chapter 3: Fight or Flight

Your rifle is your very best friend when on foreign soil. It is your interpreter, for it speaks only one language that is understood by all. But if you do not keep your rifle's singing pipes clean, it will jam. Don't be afraid to go back to the basics of Basic Training. The dust in the desert is very fine, especially in Kuwait. Invest in a barber's shaving brush to wipe fine dust away, and talk to your Supply Sergeant about getting dust caps, pipe cleaners, barrel rods and small bottles of oil. Cans of CLP (a lubricant based cleaner) are worth their weight in gold to a deployed soldier. Spend a great deal of time cleaning and lubricating your weapon before leaving this last secure zone. If you lack the discipline to keep your weapon from getting dirty, it may fail you when you need it most. With the exception of a Ka-Bar or bayonet at your side, this is the only line of defense.

When we went out on crowd control the Iraqis were killing troops by unsheathing the soldier's own blade and stabbing them right under their vest. Some terrorists are also skilled pickpockets and have fingers nimble enough to pick the pins right off your pouched grenades, and three to five seconds to get away. Sometimes it could be worse, if the enemy embraces you so that there is no chance of throwing the explosive back at them. Then they get to see Allah, or they think. So we just stopped wearing knives or frags within reach. If a close quarter fight breaks out you'll always have the butt-stock of your weapon. Always have your rifle no further than one arm's length away. *You* may have mastered this machine on the range, but make no mistake, it will obey only the one who possesses it. Always maintain muzzle awareness. The barrel of a turret gunner's rifle should usually be pointed at three o'clock when entering or exiting checkpoints unless signs or guards say otherwise. This is to prevent any accidental discharge of your weapon.

Hostages are worth a lot to the insurgents. There is nothing you can do to stop this except look more aggressive. Studies show over and over again that the scared or complacent "soft targets" are more likely to get hit or captured.

If you see a reflection in the distance, point your weapon at it. This might be a mere shack window too far off to see, or it could be the telltale glint of a sniper scope. Any loud cracking of a rifle and you'll know which one it was and where to fire back. From the time you hear the telltale crack of a far off sniper round begin to count, "One thousand-one, one thousand-two, one thousand-three..." until you hear the sound of impact nearby. Whatever number you are left with, multiply it by one thousand. Now you have distance. Call it out and engage your target. This principle will be used in the following scenario:

Private Joe Schmo is scanning his sector (looking out past his designated rifle) during his first convoy ride to Iraq. He has the twelve o'clock to the six o'clock, which means he is looking at everything on the convoy's right side, when he sees a glimmer in the distance. He doesn't freak out; he merely points his weapon in that direction. The "safety" of his rifle is still on. However, when he hears a sound like a bullwhip cracking in the distance, he immediately takes aim at the reflection and begins to count, "One thousand-one, one thousand-two..." and hears the impact of the bullet hit the embankment in front of him. Joe stops counting and yells to the squad, "Sniper, one-o'clock, two thousand meters. This is so they know where to shoot back if the sniper's next round hits Joe. (His knowledge of the sniper's whereabouts is far too precious to die with him.) Joe is now clear to engage and places his selector lever on "fire" and shoots at his target with well-aimed bursts. Because his ammunition also contains tracer rounds, his platoon can see the direction of his bullet path and can also engage the target even if they didn't hear his verbal direction and distance. When the target is destroyed, Joe places his selector lever back to the "safe" position and resumes scanning his sector. A rush of adrenaline will take place during any type of engagement. This is perfectly normal and is often referred to as the fight or flight syndrome. Your whole body will shake after the smoke clears. They rarely show that in the movies.

Just do what you think is right; every battle scenario is different and none of us have ALL the facts. When the bullets start flying, nobody knows how they are going to react. I've seen the timid lower-enlisted step up to the plate with guns blazing while their big, macho squad leaders and platoon Sergeants cower in terror or run away until the awards are passed out.

For now, don't think about what just happened. Thank God real quick for protecting your life and remain focused on the task at hand. Remember that there are other dangers possibly waiting ahead, usually sniper fire is a precursor to a roadside bomb or vise-versa.

Never underestimate the enemy. *Never.* The second and third attack is meant for over-cocky troops that just popped their cherry with a first gun battle. What we, as a well taken care of army, neglect to think about is that the enemy is used to this sloppiness of war. They've been killing since they were teenagers.

To recap, do not let your guard down even for a second when you are in the slop. Always stay Over-Vigilant when outside the gates. If you've survived the convoy trip to your permanent duty station for the rest of your tour, CONGRATULATIONS! Now you are going to have to lock and clear your weapons whenever you return from *the wire*. Every time troops enter a military post or complex they must remove all ammo from their weapons: This is a procedure that becomes very repetitions because of how many times you will do it each day. After a long mission everyone wants to go back to his or her living areas and take a nap. Soldiers who error when clearing crew and personal weapons can cause an accidental discharge of the weapon at the clearing barrels. The best-case scenario after an AD (accidental discharge) is a loss in rank and pay. The worst case is the injury or death of one of your buddies. Please pay close attention to what you are doing. Your deployment will end eventually, and then you can apply for as much prescription anti-anxiety dope as you want. Until then, stay focused because lives are at stake.

Chapter 4: Kings for a Year

When getting assigned and acquainted with your new digs, temperament among the higher-enlisted and lower officers will swell like a fresh pimple on prom night. Just let it go, these folks are under a lot more pressure to get things done on a larger scale and it should pass after a few days.

Secure anything you can from the previous group that you will be relieving. Make friends with them and see what they don't want to pack up. Stereos, palace furniture, ammo drums and/or whatever else they can't ship home goes for rock bottom prices. You can always throw it away later if you don't need it. After several days of training with the group you will be relieving, expect to start going out on patrols day and night. Our company was split up and attached to a much larger battalion in the Green Zone. The Green Zone is one of the largest military posts in Iraq, and chances are you will spend some time in a similar area when you are deployed.

Bragging is a recipe for a big crap sandwich, and dessert is usually humble pie. Our new host unit was Regular Army, so they distributed among themselves whatever few good vehicles and gear we brought and exchanged it for all the broken and unsafe equipment they had lying around. Since it was now their post that we were attached to, we could do nothing about it. They were the fancy kings and we were the surfs. But these guys hadn't trained like us so they weren't much good in the Red Zone. This is the area surrounding the Green Zone and chances are you'll spend time out here if you're hated by any top brass. My company usually was. We spent most of our time running convoys and patrols out there. Our new Artillery unit had a lot of problems with the mission because they were not trained Infantrymen as we were. These pouges started off like royalty but eventually came to us for guidance on how to conduct patrols because they were screwing up and getting a lot of people killed. It never hurts to be nice. We kept our mouths shut and helped them do their jobs when we could.

During your stay in another country, you will see a lot of wildlife that could include everything from snakes to hedgehogs. Soldiers love to tamper with them out of boredom, but resist the temptation. There may be stray dogs in the area, but these aren't your average Lassie. They've probably been feeding on human corpses outside the gates and these animals tend to carry rabies, lice and fleas. The local third world canines are pretty good at survival though, I'll give `em that. Rats are everywhere. Pay attention to the way they act, because if a large number of them come scurrying out of the sewers chances are a flood is on its way. This is a good time to get your gear to higher ground because crap water is nasty and hard to wash out.

War is no place for Germaphobes. Unfortunately, some of us are. Personally, I would have rather dug a cat hole every time I had to do business, but sometimes it's impractical. Say goodbye to indoor bathrooms for a while unless you get a cushy office or supply job. If you are combat arms, you will usually be doing back duty in overly-used port-a-potties. I would take a newspaper in the can with me and drop the sports page into the bowl. This act made me feel like I was a two-hundred-pound canary in a cage. But the impact of a large dump doesn't splash toilet water all over your butt. With any luck, people back home might frequently send you wet naps and it is always a good idea to keep a stack in your cargo pocket to freshen up after an ordeal like that.

Chapter 5: Riding the Trigger

One thing the Warsaw Pact areas need to recognize about picking a war with NATO Forces, we are the law when Congress sends us. Everyone answers to the military in combat because it's *our show*. We'll rebuild the country, but follow our rules first. National police act like big shots to the general public, but if you bark an order at them watch how quickly they scramble around to do it. Traffic must obey the rules of all of our convoys. On the freeways of this country there are no rules except the end of a barrel. When entering a "freeway" on-ramp, blow your whistle at the flow of traffic as you merge. The result should look like what happens in the states when an ambulance goes by with lights and sirens. If any should feel like he doesn't want to stop, point your weapon directly at him. If this isn't a deterrent, point weapons at the side of the road and let off a few warning shots.

What goes up must come down is the old rule. Never fire warning shots into the air! Your pissing contest with some arrogant road-rager could end up with a child a mile away with a limp for the rest of his life. Always fire warning shots into dirt, not asphalt, as they could ricochet. If this driver still persists on entering your convoy or trying to go around without you clearing and waving him through, he is now a threat. Go for the engine block first. If this does not stop the driver, walk your rounds right up the hood of his car until this threat is eliminated.

After any combat engagement it is sensible to tell your boss: "I felt my life and the lives of the convoy were in danger." That should go into the report and clear you of disciplinary action, if legitimate. Of course, this will not clear your conscious if you just felt like shooting him.

Many insurgents will plant explosives in their cars and detonate them as soon as they are close enough to a convoy. The other irrational drivers pose no threat to the convoy and are just mad enough to break rules because we invaded their country.

With great power comes great responsibility. You have to make the call on your own. I've known troops that double tapped an innocent driver and still have nightmares about it. I have also known of soldiers that hesitated and are now in heaven. Convoy gunning is the most difficult job a person can volunteer for. There is a lot of gray that goes with this position. It's never black and white. With this job you have to play judge, jury and executioner when necessary. You hold the fate of all the passengers of your convoy in your hands, as well as the innocent majority of Nationals. But don't let this go to your head. I've seen GI's get into trouble when they abuse this power. I saw a troop shoot out a car window because he was having a bad hair day. Do not let this be you. The guy lost rank and a cut in pay over his little hissy-fit. The ones that really screw up their gunning jobs wind up dead or guilt ridden. But hey, this is war and it ain't pretty. Just remember that everything you do in war is means to an end.

My Company Commander, Captain F, used to repeat this statement to us: "If it's them or us, make sure it's them."

Our company mowed down anyone who posed a real threat because we knew it was a necessary means to seeing our families again. We would allow nothing in our convoys and patrols. In combat, your brain slows everything down, but the trigger of your rifle is the fast-forward button.

Make sure your target is a confirmed threat and use well-aimed shots.

Provided you survive your firefight, you will need to give your superior a quick report on the situation. This is called a Situation Report; or *Sit Rep* for short. When your Sergeant or Lieutenant comes by, he will ask if you are hurt, need ammo, water or a medic. Tell him you are "up" to each question if you don't need assistance so that he may check on his other guys. Keep scanning your sectors of fire. Just because the first attack has been dealt with does not mean there won't be a second or third ambush waiting. Usually the first attack is to throw you off guard for the real battle. Make a quick mental note of what just happened. You will be asked after returning to post what exactly went down, but you can never go wrong with "I felt my life (or the lives of my crew) was/were in danger because of ..." and tell your story.

Talking to a Chaplain afterwards always made my platoon feel better and I recommend it.

Chapter 6: Press

What can I say about the press? My experiences were usually pretty good, although you generally have to be very careful of what you say to them. As you may or may not know, the press usually gets to go wherever they want to, with the most professional army in all of history watching their backs. They will give you smokes when you need 'em, joke with you about what a mean jerk your platoon Sergeant is, and take *really* good notes on what you think about the war. Then after about two months or whenever they have their big story, these reporters catch the next hop flight right back to the states. Don't say anything to the press that you do not want in print.

"Off the record" is a bunch of crap. The tape recorders are always running. Some of these guys already have the angle or scoop they want before they even arrive at your base. Something you say might tie in great to the reporter's story, and leave you with a court martial while the writer you confided in is clearing space on his mantle for his next Pulitzer Prize.

Classified means classified.

An easy rule to keep you from saying too much about military operations to the media is: If you are not in charge of it, don't talk about it. For instance, I was in charge of an M-203 Grenade Launcher and an M-249 Machine-Gun and that's all I ever talked to the press about. There is a lot to know about those guns, and it's all been declassified. I would rattle on and on about the cyclic rate of fire and how well they were made. This showed them that I was a squared-away troop that knew my job. That's it. Never give the Press anything that could possibly compromise a mission or other soldier's lives. Whenever the reporter would shift the conversation on something I felt was privileged information, I would change the subject and talk about God or how much I missed my family. If anyone snaps a picture or records something that they shouldn't have, my advice would be to seize the article and report it to your chain of command. We don't need any more America-bashing, and it's bad for troop morale. So what if a paparazzi loses his story.

My intention of this chapter is not to make you overly paranoid. These men and women are doing a job and most of them do it well. In fact, I still keep tabs on a buddy at the *Times* who always tells a story the way it happens and didn't once misquote me, so there are some cool field reporters. Just talk to these gents as if they were your sister-in-law. Be nice, talk about the weather, but keep them in the dark about any juicy gossip you may know. It's none of their business. You will have plenty of time to talk to the press when you have finished your tour and are back in the states. Then you can dish the dirt and have an opinion. But until the uniform comes off permanently, you are Government Property and subject to UCMJ.

Chapter 7: Pouges and Rouges

A lot of the soldiers deploying to Iraq will spend their tour in the Green Zone as specialists in the field of purifying water, fueling vehicles or administration as if they were on an ordinary post in Anytown, USA. Some of these men and women conduct themselves in a very professional manner and I am not addressing them. I'm talking about the folks that do their military occupation in a sloppy way, not even knowing how to defend themselves with a rifle if it came down to it.

This is wrong!

These folks are what the Infantry and other Combat Arms specialists refer to as pouges. They really don't do diddly. They usually pick a safe military job before going to war just for the shiny badges and bragging rights of being in the military.

Iraq's Green Zone may look comfortable deep within the core of the area, but strip away all the razor tape, towers, cameras and jersey barriers, and there is nothing to keep insurgents out except personal weapons, which many of these pouges choose not to carry. Many times, my platoon would be strolling back through the Green Zone gates and see some *butch* jogging in sweats without her weapon. Female American soldiers are the most sought after hostages in all of the Middle East because the local male populous does not believe in granting a woman the same rights as men. When Middle Eastern people see these ladies filling roles that provide them with authority and "big guns" it enrages the Iraqi males. When unarmed females are captured from easily accessible areas of the Green Zone and other coalition bases throughout Iraq, they are usually taken prisoner, beaten, raped and beheaded. It is an unfortunate and tragic part of war that is hardly mentioned in the media anymore. These women should better arm themselves and realize where they are.

For the men, don't ever take a piss anywhere that a female might see. This urination includes out in the desert. It really has become sissy wars in many respects. If a soldier girl happens to see your stuff, she can report you to higher-ups if she feels like it.

Then guess what?

You lose rank!

On the other side of the spectrum are the Delta Forces. I really can't say too much about these rouges, because a lot of what they do in Iraq and Afghanistan is still classified. We escorted a lot of them through the red zone, so in essence; our company was their "protection" in certain respects from point Alpha to Bravo.

What can I say, those missions kinda made us feel like big shots.

Combat Arms is a good place to start if you are interested in obtaining *The Tab* and upwards through a *Tower of Power*. In Special Operations, you've got about a year or so of language, survival and weapons training before they send you out on the real jobs. But with a re-enlistment bonus currently at six figures for these operatives (along with tax free base pay, combat duty pay and a whole lot of other perks going for the tuffs who end up touring the world's hot spots) it's worth checking into if you really want to live life on the edge. From that status, the shadowy Deltas may call on you.

Another route is working for independent contractors such as Blackwater. These private security firms provide competitive salaries to prior military personnel, but it's a good idea to have a tour of duty in a war zone under your belt for them to even consider your resume. Also understand that when joining a civilian-contracting group, you will not have the protection (calling for an airstrike, for example) or as much comradely of the military watching your back, because you no longer belong to them—you're a Mercenary.

There are many different roads to travel in the service. But with recruiting at an all time low due to the politics involved in the wars of Iraq and Afghanistan, we may yet see another draft. It is unfortunate that so many Americans love living in the United States but as soon as she calls for help, many of our able-bodied youth threaten to run to Canada.

If you are reading this as a service member—regardless of whether you are man or woman, black, white, Chinese, Hispanic, Rouge, Pouge, Combat, Non-Combat or any other difference, you are a part of something big. You are the "Jolly Green Giants" walking the Earth until you return home. God bless you for what you are doing for our country.

Chapter 8: Traitors

Here's the ultimate slap in the face:
Not only do we employ certain "loyalist" Iraqis to do our dirty work, i.e., cleaning out the crappers on post, emptying garbage cans and other menial tasks for a moderately handsome salary; many of these back-stabbers walk around each base stealing things they can sell to the insurgents. This includes information on where our ammo points are, maps, pace counts and addresses carelessly thrown in dumpsters by complacent troops.

Always burn sensitive mail!

Make sure that any mail you receive is kept in a safe place, remove all addresses and promptly burn them. There are rumors that insurgents send mail bombs to these addresses. That's the last thing anyone needs to worry about when fighting a war, so take an extra minute after reading your mail and torch the envelope it came in. Trust no one. If a national is acting weird, he's probably up to no good. Don't be afraid to keep the suspect detained and report your findings to higher-ups. Traitors steal military secrets from posts on a constant basis; even the most mundane scrap of paper could be a valuable bit of knowledge for hostile forces. That type of leak gets the wrong people killed. It is your job to watch out for clues as to who the traitors are while abroad.

Chapter 9: Pissing Contests

There are other soldiers that are going to try and screw you over, and this is especially true of soldiers from other units, and all types of ranks. The only people anyone really cares about in combat are themselves and their buddies. It's very primal and animalistic, but hey, that's war. So if you are fueling up on another post or eating in an unfamiliar chow hall, just remember to keep your mouth shut and watch your buddy's back. Same also goes for other guys and gals misbehaving on your turf. Be polite, but if someone is doing something on your post that violates your unit's S.O.P. (Standard Operating Procedure) be sure to tell them so. When the soldier outranks you, always address them by their title before and after instruction of any kind. This way you won't have some power-hungry non-com with too much rank writing down your last name. On posts you really have to be careful who sees you do what. You can bend and even break the rules—just don't get caught.

Always stay awake on guard duty.

This is one of those pointless dog and pony shows during peacetime that could prove invaluable during war. Don't be copping any Z's during your shift. If you are lucky, you will wake up with the guard list in your hand at dawn and realize you fell asleep without waking the next soldier on your rotation. A not-so-lucky troop caught napping may be killed by the enemy and never wake up.

Geneva Convention—friend or foe?

Nobody really knows anymore. The enemy gets to use whatever means necessary to keep us out of their back yards with their Warsaw pacts (Basically, no rules) whereas the United Nations abide by many rules on the battlefield. Every week the socialists cripple military tactics a little more in Congress as the Warsaw countries plant mines and torture captives. But that is what makes us Supermen. Even with our hands tied with kryptonite, we are the Elite, Kicking and Biting Justice League of the world.

Chapter 10: Emergency Hypnosis

For centuries, man has been exploring the inner recesses of the mind. It unlocks the part of the mind that became the focus of Dr. Sigmund Freud's lifelong work. Right up until present day, hypnosis has been a topic of psychology that has had great controversy among many experts of the human brain.
When Morphine is not as available on the battlefield, emergency hypnosis could be an effective last-ditch effort to staunch blood flow and relieve shock. Magicians use this talent to astonish crowds, and therapists use it to treat patients with addictions.
To fully understand the methods of this technique, you must understand that we are all suggestible to hypnosis in one form or another. Every day our senses send messages to our brains. Taste, touch, sight, sound and smell are receptors that feed directly into our conscious minds. These *message units* are sorted by the critical mind, which acts as a barrier to the unconscious mind.

There are two factors that motivate hypnosis:

Pleasure, which is a *known* to us, versus pain, which is an *unknown*.

As human beings, we are drawn to pleasures of life, even if these pleasures are self-destructive. However, change is an unknown. This situation is not easily taken in by the critical mind. Hence the reason this part of the mind needs to be temporarily neutralized in order to manipulate the subconscious. Anyone who claims to induce hypnosis through relaxation has not learned the concept of this skill. Hypnosis is brought about through stress of the subject initially. Too many message units swarm the conscious as someone is induced and result in *Fight or Flight:* an overload of the critical mind. Fight or Flight was a survival tool granted to us by our Creator, to either outfight or outrun dangerous predators of ancient times. Since we have little use for this feature in the modern world, fight has become anxiety (literally suppressing rage) and flight has become depression (literally escaping from ones problems, if only in one's mind) when in this state of heightened awareness, the critical mind breaks down temporarily.

Now the unconscious mind is accessible and recordable. This very fragile area of the brain has no sense of reason. Basically, the unconscious mind is where beatings from high school reside, or the time that someone called you something that really left an impression. Our self-esteem and traumas are kept here, according to most theories.

The subconscious is a direct and literal attribute. Tell it to do something and it will. Access into this area for reasons of reprogramming is the primary goal of the hypnotist. In order to do this, the therapist must first create an overload of message units and subtly "stress out" the patient before proceeding to reprogram as a means of temporarily calming the patient on the battlefield. Advertisers have studied this method well, and most television commercials suggest a stressful situation before offering their product to alleviate pain.

The only factor that makes the job of the hypnotist harder is when a subject refuses to take the hypnotic suggestions seriously. This will not be a problem for the wounded on the battlefield. They are already in a state of shock.

Note:

Before attempting hypnosis, make sure there is no other option to relieve the screaming outbursts and pain of your wounded. Hypnosis is also only to be used if medics or anesthetic are not available. Hypnosis can possibly be emotionally scarring if performed incorrectly. But hey, so can war. Use this information at your own discretion and risk as well as the risk and discretion of those around you. Make sure to follow military training for field dressing, tourniquet (if needed), shock, etc., and make sure someone is pulling security for your location before you begin. Now you are ready. Have the patient focus on something in front of him/her and say in a confident, even tone that:

"You will only respond to the tone of my voice as you slip into a state of deep relaxation. When I count down from five you will be fast asleep as your body begins to heal itself. Four. Everything your body is doing at this moment and continually is sustaining your healthy life. Expect a full recovery from any wounds you may currently have. Three. There is no pain. Two. Any open wounds on your body have already clotted up and are beginning the healing process. One. (Snap your fingers close to patient's forehead and lightly touch between his or her eyebrows. This stuns the conscious mind.) DEEP ASLEEP! You are conserving all of your strength to enable the healing process of your body. When I count from one to five, you will wake up completely refreshed, relaxed and healed. But for now, sleep."

Now you can leave this soldier in a state of hypnosis until help arrives. Then make sure to bring him/her out of the trance:

"I am going to count to five. When I reach five, you will be awake and may take all direction from your doctors. One. You are becoming more aware of your surroundings. Two. Every part of your body is functioning the way it should be. Three. Your body is responding well to all lifesaving treatments, and your recovery time will be quick and complete. Four. You want to live, you have a lot to live for, and you feel a new sense of purpose. Five. (Snap fingers again and touch brows as mentioned above.) "Wide Awake! Wide Awake!"

<u>Again, this is only to be used in an emergency situation when all other methods of treatment are ineffective or no longer available.</u>

Chapter 11: The Art of War

Before each mission, you will go through preliminary checks with all of your gear, weapons, ammunition, rations and water. Some of these checks will take place at the staging area where you are to meet your liaisons that need protection for various types of missions. These men and women are usually top brass, and are usually called *Principles.*

Make sure you are squared away before entering the staging area. This includes a clean rifle, supplies and a working knowledge of what you will be doing for each particular mission. The last thing you need is for your squad leader to put a boot up your butt for forgetting something vital, like batteries for your night vision goggles or something that sounds petty enough to forget and need later.

When you leave the staging area, that's it.

Your platoon will be leaving the safety of a post and into the chaos of uncertainty. *Incoming* is the scariest word you will hear in a combat zone. Explosions are the norm in the Middle East, and you will get used to them in time. But when "Incoming" is yelled and heard before or following explosions, dive under something solid, real quick. What this word means is that the enemy is firing mortars, rockets or any other serious killing ordinance in your direct vicinity, and occasionally with accurate results.

You will not always hear the mortar's whistle (that is for the movies) but you will see and hear the impact. This is when the world moves in slow motion, and everyone is scrambling around for cover. As soon as you find yours, take a look at where the explosions are landing. If there is a radio nearby, use it to call in your position and for a medi-vac if needed. There will probably be casualties, especially in a highly-populated area. Try to determine what direction the rounds are being fired from, and call in that location as best as you can. That way any armed aircraft in the area can look for the heat traces of a mortar tube and shake up a few grid squares.

Whatever happens, do not leave your buddies behind bleeding to death.

Inaction will haunt you for the rest of your life if there was a chance to save some of them. Any effort to do the right thing is commendable. If you do happen to freeze up, try to let it go afterwards. We are all only human, after all, and <u>no one</u> knows how they will react to real world incoming for the first time. Many soldiers piss their pants after an ordeal like that. I certainly did, but the stain isn't too noticeable thanks to our uniform's intricate camouflage design. Believe me, that is the least of your worries. A quick prayer works wonders, as you stay vigilant while being attacked. The good news is that mortar volleys usually don't last more than a few minutes, thanks to our high-tech air support flying around and ready at a moment's notice.

Chapter 12: Bootleggers 'R Us

Try to imagine every DVD known to man in one country the size of California. Iraqi bootleggers will sell you all the new releases on DVD during your tour before the films even show in theaters back home. Provided, the opening credits are written in French and you can see small heads bobbing around at the bottom of the screen as if you were in a virtual movie theater. On occasion, someone from the audience where this contraband was filmed will get up for a bathroom break.

The way to buy bootlegs in Iraq is to try to look for titles that are at least two years in circulation. These are usually stolen from computer downloads by locals and they do an impeccable job at $3 a pop. If you fancy the new releases, be prepared for crappy quality from a download traced back to a recording device secretly planted in French theaters.

What does the Army have to say about bootleg purchases in Iraq? "Enjoy yourselves!" After all, this is a war and the military has a lot bigger things to worry about than soldiers paying a fraction of the cost of a DVD at the Post Exchange by supporting the local populous. A portable DVD player with a small rugged case is a good investment for your tour and can be bought at any Post Exchange. You might also want to purchase a binder-sized CD case and small foot locker (with lock) to keep the whole mess secure. Just make sure you only have only one of each DVD title when you leave Iraq. Otherwise, customs will confiscate the duplicates. They don't want you going home selling them in bulk. They also will not let you bring back porn of any kind, so try not to invest in too much of it over there.

Just so you know the magnitude of the situation here, the continent you are rebuilding is supposedly chock-full of ghosts. I don't know if this is true, but one thing I am sure of:

Iraq is inhabited by dark forces, also known as demons. Between the Tigris and Euphrates Rivers, the Garden of Eden was swallowed up because of a fallen angel known as Lucifer. This evil being took the form of a snake and tempted Eve.

I think this could be the area from which Lucifer was cast from Heaven. I believe this area to be one of the possible locations spoken of in the Bible's Book of Revelations regarding the Apocalypse, or end of days as we know them. Anyone allowing the mark of the beast (probably a microchip) to be implanted on their right hand or forehead will be damned forever.

So it is my theory that this country is where a multitude of demons still hang out. They love war, and the only protection against these foul creatures is God Almighty. I saw a demon following our platoon as it possessed large dogs that fought to the death as our armored escort rolled through the ancient city of Babylon. As we made our way to the next town, a small boy was the demon's puppet, and finally an older gentleman smoking a long pipe and staring at me with the entirety of his eyes a color darker than glazed black onyx.

There was nothing the demon could do because I have the Salvation of Jesus Christ that protected me from whatever could have happened if I was not Saved by Grace. There were also countless times that my platoon and I saw and heard apparitions in many forms, even when we laid off the hashish:

Men and women, streaks of bizarre lights, and inhuman sounds. Not to mention the clouded evil of the living insurgents on a day-to-day basis. There is an unspeakable evil that makes these human monsters operate in such a cruel manner, claiming it is for Allah's glory. What separates us from these bastards is that they will think nothing of using a child as a human shield.

Anyway, just be careful out there and if you have any more questions on the subject, there is always a Chaplain at the ready with answers to life's toughest questions.

Chapter 13: Understanding Terrorism

NATO and Warsaw pacts are the rules of war. The United States, as well as most of the other civilized countries, belongs to NATO. This set of rules is implored with the design of our weapon systems. For instance, the barrels of our weapons are spiral bored. The reason is to give each bullet a spin in flight, ensuring a clean exit wound. Many countries in the Middle East, as well as North Korea, belong to Warsaw. If you get hit with one of their bullets it is a lot messier, especially if the round hits bone. This turns the projectile into a sort of pinball. Warsaw still plants unobserved mines, whereas we manually fire claymores in order to know what and who we are blowing up. Even though it is more dangerous to hang around after blowing off one of ours, at least we know what our targets are.

Let's say while you are in the Middle East you engage in a firefight with a few enemies. These people are going to leave behind about ten children to fend for themselves. One day these kids will be rummaging through mom and dad's things packed away in the closet, and find propaganda with a few stashed AK-47s. A decade later, these grown orphans are the new threat. By killing one insurgent you have inadvertently created ten more of the next generation by killing their parent. That is why we go to war with the Middle East every generation or so, it takes that long for a new batch of terrorists to rise up against us with enough pure hatred to plot such diabolical schemes as the World Trade Center. But you are there to do a job, dear soldier, and these people started trouble with us first. Do the best you can to show the local kids we just want freedom and peace for all.

The United States really tries to wipe the butts of all the other countries of the world. Sometimes we get crap on our hand. Every time I felt like I should not be in their back yard, I would remind myself of those burning buildings in 2001. That thought kept me from getting all sappy over there and made me keep my edge.

All the celebrities that speak out against this war should leave the United States. "Love it or leave it" as the motto goes.

No one loves a soldier until the wolves are at the door. The reason our country is so great is because of our constitution, which allows these rallies to take place in the first place. I'd like to see our protesters in another country protesting about something they don't like. They would be shot. This is why terrorists from third world countries hate us so much, we can express our opinions in America, and there is no other place like it. Due to the sacrifices you are now making for us now, dear soldier. So be proud of yourself.

Chapter 14: Support System

The only thing that makes war bearable is Almighty God, the people you have back home rooting for you and the friends you make throughout your journeys during the time spent in a combat zone.

Your company is like a large family. The commander is like your mother. Everything is etiquette in his or her household. The mother of this family has lieutenants under her, your older sisters if you will, who will rat you out in a minute if you forget to wear your booney cap outside or neglect to clean your weapon. These fancy soldiers went to the same proper schools as mom, and much like daughters, they have high hopes of being mothers to their own companies someday.

Now mom always worries about what the rest of the neighborhood (rival companies) think about her household. Every time her kids misbehave (that's you), it reflects directly on her. This results in disciplinary action of her children, which is usually handled by the "dad" in charge. Unlike the commissioned officers, a First Sergeant had to work his way up through the ranks from lowly private. Even though he is on the same pay grade as the commander, mom's word is ultimately law and dad has to enforce all whuppin'.

Under the First Sergeant are other Sergeants. These are your older brothers. These guys will slap you around and keep you in check to prevent a huge grounding from dad. Sometimes your older siblings can and will abuse you because they forgot what it was like to be the younger brother. That's why you have siblings known as battle buddies to watch your back. You will do everything from play poker to clean the crappers with these fellows. Most will live and some might die, so always watch their backs as well as they watch yours.

There is only one rule to keep your battle buddies good mates:

Never screw them over. If a situation arises and someone has to come forward and take the blame, do it. Chances are, you will be respected every single time and gain a reputation among subordinates and superiors alike of matchless integrity.

Chapter 15: Everything but the Sink

This chapter is sort of all over the board, but very important to your survival, so I decided to leave it in this report. The topics of this chapter are anything of value I may have touched on in other chapters, or may have inadvertently left out so far.

Insurgents will use every devious trick imaginable to make you feel unsafe and start to doubt yourself, including propaganda leaflets designed to brainwash and confuse you. There was a time when a photograph of a hostage by the name of "Jon Adam" who was taken from our post turned up. We had accountability data to show that no one was missing, and yet here was a photograph circulated on CNN of a soldier in full battle rattle, bound in front of an Arabic banner with a carb 4 pointed at his head. Our Intel Agents discovered that this "troop" was a 12 inch plastic *GI Joe* action figure by the name of

Agent Cody, purchased at the post exchange by a civilian national and sold to the insurgency for their scaled up photo shoot.

If that's not bad enough, we have so-called Americans such as Michael Moore living off the fat of the land in America, making propaganda more harmful to American soldiers than any enemy of our country could create. This war has become a political mess because of people like Barbara Streisand and Jane Fonda. Many other Hollywood celebrities and college professors are doing a lot of damage to the credibility of the American Soldier. College lifers and Lefties enjoy subtly making us look like baby killers to the public at large instead of screaming it at us, for now.

The views of a portion of the American public are a tamer version of Vietnam in many ways, but don't you believe a word of it. "No one loves a soldier until the wolves are at the back door" rings truer now as much as any other American war. If we allowed terrorists to get away with these atrocious acts without calling their bluff, they would do it again. Any sympathizers of these insurgents know now that we will never forget September 11th. We may forgive, but never forget. Keep your head up and don't let these arrogant American Liberals keep you from doing your job overseas.

Water is a precious commodity in the Middle East. I may have said this before, but repetition is important. Keep yourself very hydrated. At least one 8-ounce bottle of water an hour.

One of the most asked-for commodities by American GI's of both genders is Hosiery. Not only do they wick away moisture and revitalize sore muscles, pantyhose keep the sand fleas from crawling up trousers and biting you where it counts.

Chapter 16: Going Home

As your tour of duty in war grows short, it will be very tempting to get the *Short Timers Syndrome*. The symptom of this particular mood is complacency.

Complacency kills!

You'll have plenty of time to call up old girlfriends when your boots touch American soil once again. But your attention until the very last minute in a combat zone belongs *in* the combat zone. Otherwise you will end up dead, or worse, getting one of your buddies killed. Too many times did I see short-timers with a few weeks, or even *days,* in-country get their heads blown off. Please don't let this happen to you.

Information to keep.

If you don't have a 3-ring binder of some sort, buy one at the Post Exchange and file away your important papers, including award paperwork, Department of Defense forms and *all* of your orders—old and new. That packet is your ticket off the *strange little Middle Eastern planet* and back to America, so keep it somewhere safe. Every time your superiors give you any other paperwork of importance, just put it in the folder. Also keep all the gear on your exit packing list. Everything else is your choice to carry or get rid of. But if a piece of gear is on that list and you don't have it when you get home, be prepared to buy it from the government at what they paid for it:

Full retail!

Information to destroy.

Burn any sensitive material you no longer need. Addresses, classified battle reports, maps, memos and anything that no longer has any value to your journey back home should be placed in the company burn barrel and torched. Use your own discretion. I was so paranoid out there that I burned my used Q-tips and tee shirts just so insurgents would not have access to my DNA, but that might have been a little ridiculous. On the other hand, I may have mentioned that we've caught the nationals rummaging through the trash with crumpled or torn pages of classified data in their grubby little pockets.

Sell it or Send it.

Start streamlining your gear as much as your chain of command will allow. The more you get rid of, the less you will have to worry about. During your tour, you may have received piles of letters. These need to be boxed up and mailed home. All the DVDs bought in country should be stripped from their bulky envelopes and sent home in a DVD carrier.

Be prepared for some ridiculous crap.

There was one jackass squad leader I knew of that actually made his troops polish their remaining bullets for turn in to the next soldiers coming to relieve us. The Army can be a corn dog and pony show due to a few arrogant decision makers in this otherwise outstanding organization.

Take pictures.

The non-classified items you are allowed to photograph should be snapped up, because even the dull and mundane in another country can be quite fascinating to friends who never left your home town.

Tell the unit who relieves yours EVERYTHING you learned.

Just as I have shared my experiences with you, the reader, now it is time to share anything of informational value to the new cherries. Anything they wish to know about their new tour of duty and everything of importance that helped keep you alive might also be valuable to their survival. Swap stories. If it weren't for this new unit's rotation, you and your crew might have been stuck a while longer. Be grateful to them!

The process of going home is a long one.

They call the next part of your experience *Ku-Wait* for a reason. Just try to be patient with your bosses and the lower enlisted you outrank. Everyone is trying to achieve the same goal—getting the heck out of there. Now is the time to start gathering all the paperwork from your awards, medical visits during your tour, anything of importance. Try to make copies and file them away in your binder. If you miss this chance, there should still be an opportunity to obtain these papers when safely de-mobilizing in the United States.

Do not transport ANYTHING that will delay the Company's return.

This includes hookahs, illegal drugs, unauthorized and non-issue weapons, ammunition of all sorts (you will probably recycle ammunition stockpiles to the unit who relieved you anyway) live or inert ordinance of any kind (especially grenades!) blasting caps, composition 4, questionably large denominations of currency, two or more bootlegged DVDs or CDs of the same title, expended ammunition casings or links, swords, anything made of an endangered animal, any insect, any human body part, war trophy, artifact of historical significance, literature promoting enemies of the United States, maps of your tour, lewd photographs and pornography.

There are amnesty boxes everywhere. These are really your last chance to get rid of contraband before a thorough search. Just make sure not to put ammunition or ordinance in these boxes (it could endanger lives). Any live or inert explosive or ammo should be handed over to one of the cadre before the search. Just tell them you found it nearby and wanted to do the right thing.

A general rule and then we'll move on.

In the military, as long as you come clean about it *before getting busted,* you should be okay in most circumstances. This applies to everything from a drinking problem to contraband shake-downs. Just don't wait until the last minute. It ain't worth it, so just get rid of it. Whatever you are thinking about smuggling, about a thousand guys or girls before you tried the same thing and got detained. Then these troops received a dishonorable discharge after all that drama of war they went through. The only job those MPs in Kuwait have is to bust war heroes out of jealousy. Don't let this happen to you.

The long journey home.

This is where soldiers get unnecessarily sloppy, so I will spend the remainder of the book discussing how to blend into society once again. Don't get drunk *on the* jet ride home. There will be plenty of time to celebrate. I made the mistake of getting a whole lot of booze when our chartered 747 landed for a pit stop in Ireland. This was not smart. Not only did I offend half the plane with drunken antics and blatantly true remarks, but I nearly lost rank. Keep it clean until your flight is over and you are back on American soil.

YOU DID IT!
YOU ARE HOME!!!
Now more waiting.

In the military they have a saying, "hurry up and wait." This phrase will never ring truer than what lies ahead of you for about two more weeks.

You will land in an international airport, and hopefully walk through the terminal to a loud, welcoming cheer. If this is not the case, *so what!* The politics of this war are a fraction compared to those of Vietnam, but it is a bit of a crap shoot what kind of greeting you will receive. Most civilians will never know what kind of hell you went through for their freedom, so good response or bad, just laugh it off. Foul opinions towards the troops are as worthless as the people who give them. Ask these folks what they did for their country lately and that will usually shut them up. Especially those worthless college twerps that want to run to Canada if the draft comes back.

From the airport, you will be transported to a military post. This is where you will de-brief and go through gear turn-in and a slew of mental and physical check-ups to update your medical and dental records.

For this last physical, be BRUTALLY honest.

This is the time when you describe every wartime injury in great detail. If you saw a lot of terrible things in war, tell your shrink. If you sustained any back, knee, head or shoulder injuries during your tour, let them know. These types of injuries have a habit of coming back in a soldier's old age. Let them know of any radiation exposure such as depleted Uranium.

This time at out-processing will be your LAST CHANCE to establish a paper trail on any wartime ailments. If the military does not have your injuries on paper, the injuries do not exist. Furthermore, it would be unwise to waive your physical in order to have more time on post to screw around. You are given this time to have your body checked out at no charge. Have them look at anything you feel does not work the way it used to.

Be thorough with your check-ups, but move diligently to each station. Then make copies of any written reports, and your entire treatment record. Next, file all of it away in your personal binder. Do not lie or over exaggerate with your answers, just be honest. Each medical station must sign you off as a "GO" or you will be stuck on post while your buddies catch the bus home at the end of the allotted time. Use your best judgment when speaking to the medics and communicate scheduling with your unit. There should be vans to take you to each appointment. Listen carefully at each briefing.

At night, eat, drink and be merry.

Surviving a war is something to celebrate, yet be somewhat responsible! The chain of command might cut you loose each night, but don't get stupid. There are a lot of people in your hometown waiting for you and what a bummer it would be to hear you got killed from a night on the town with your drunken buddies near some Army post? Also, you could be piss tested until the very last day you are released to your loved ones, so lay off the el dopo and make sure there is a designated driver when going out to the clubs.

Married soldiers, try to keep it in your pants.

Every married guy I know that ended up cheating with some two-bit whore after a long deployment felt guilty about it when reunited with his wife. Go home with a clear conscious; no sowing-wild-oats fling is worth explaining.

During your time so near yet so far away from home, some sound advice:

Drink beers in the barracks (where you are *finally* safe from enemy artillery) and write your memoirs of the experience you had in a combat zone. Writing is a good outlet. Besides, those clubs around post are nothing but trouble.

And now, the longest bus ride of your life.

After finishing your medical processing and equipment turn in, you should have a checklist that has been signed off. If you do not have EVERY SINGLE item checked off by the deadline given, why don't you read this book a few more times, because the finished soldiers will be leaving without you. Assuming you are on the long bus ride home, get ready for a big party at your original headquarters. Loved ones will be waiting. As you leave the bus, the Company will march out and fall in to platoon formation (or something similar). Then these long-awaited words will be shouted at the top of your Company Commanders lungs:

"DISMISSED!"

From then on, the only war you have left to fight is with yourself.

Chapter 17: Over-Vigilance

When I returned home everything spooked me. From garbage men clanging cans to fireworks, I was one seriously jumpy monkey.

I found it hard being back in America and trusting anyone of Middle Eastern decent after their relatives tried killing us over and over. This is wrong and was dealt with in therapy.

Anyway, I finally shook off the hate I felt towards the Muslims, but still find some of their abusive ways a little bit creepy. Small steps.

If you need therapy, the VA provides it free. Look them up online, or go through your out-processing packets to find the nearest psychologist to help. The occurrences you may have seen in the war are very traumatic; certainly not something a human being is meant to experience. These scars will take time to work their way out of your system, but that time will come. Better to get it out sooner than later. Some of my Vietnam-era friends are still walking around with mental baggage. The VA wants to help you get rid of it now.

Chapter 18: Chaplain's Corner

During your stay in such a horrible war zone of death and suffering, one or more demons may have latched onto you. Sometimes the truth can be scarier than fiction, if you are misinformed. The devil and his demons are very real, and they try to wage war with every person alive to keep us from learning about the Salvation of Jesus. HE is the only way to the Father (God) and by asking for Jesus' forgiveness and acknowledging His death on the cross for your sins, you will have a place in heaven—guaranteed.

Being exposed to war is nothing compared to the pain and suffering of what hell will be like. God gave us a way to prevent this from happening to ourselves, and all we have to do is repent of our wrongdoings in this life.

Christ's Gift of Salvation is FREE, no matter what we've done here on Earth.

If you want to make the most important decision of your entire existence before it's too late, say this:

"Lord Jesus, I am a filthy sinner and I call you into my heart to cleanse me of all sin. I thank You for dying for my sin on a cross 2,000 years ago (John 3:16) and I love You now and forever, my Savior, with all my heart. Amen."

If you really meant that prayer, you are forgiven and will live in eternity with the Living God. This repentance also protects you from this world, but you need to know how to reject demonic influence. If you were exposed to horrors in the war or anywhere else, especially without the protection of Jesus, you must cast out these demons in His name.

Now that you have said the sinner's prayer, God has given you the power to say this with Authority: "In the name of the Lord Jesus Christ, I rebuke any unclean demons from my mind, body and spirit. I cast out any evil demons and condemn them to water-less places in Christ's Name." Make sure to sound off like you've got a pair, and show these foul creatures you will not tolerate their torment any longer. The reason you want to cast these demons into water-less places is because the human body is made up of 80% water and it's a comfortable place for them. You must verbally cast them to a water-less place so that they aren't able to cling to any other living host.

Most importantly, you must cast them out in Jesus name. This is the only One who can rid you of your demons. Then pray to God:

"Thank You, Father God. In Jesus' name and by His stripes (Isaiah 53:5) I am healed, Amen." Jesus received a whipping (stripes across his entire body) before His Crucifixion, and with His Grace, you are now healed.

Now that you have done this, the devil will make every effort to steal you back. The good news is, he can't! Jesus now has your back, front and sides (both in and out), and you are one of His soldiers now and forever. Pray and get a Bible. Ask Him for clarity as you read. All the answers are in that Book. (Author's note: If you read *Chaplain's Corner* from the prompt in the introduction, go back to the beginning of this report as you are now protected by the Hand of God. Should you be killed in battle, He will receive you because you just received Him.)

Chapter 19: Reuniting with Family

After the whirlwind of sex with your wife or girlfriend, you may notice things around the house may be a little different than when you left. Make sure to compliment everyone on remodeling or decor changes. If the woman in your life did something new with her hair or nails, bought some new clothes—whatever, tell them they look great. These heroines kept things together despite a lot of pain, suffering, praying and worrying about you while you were away. They deserve to have you gush over them for a while.

With a little of your hard-earned combat pay (assuming you didn't squander it away over there on bootlegged DVDs and other knick-knacks) take your better half shopping at her favorite department store and dinner as many times as you can afford.

The kids.

My ex-wife and I didn't have children yet, but we did have two little dogs. The boy dog was daddy's little helper before I left to war. For months after my return, he growled when I got in bed next to his mom. Man's best friend would piss and crap systematically in front of my closet every morning to vent his abandonment issues. This was the dog's only way of communicating his anger with my leaving for so long. Just imagine what a human child with these issues is capable of. If you have kids that now misbehave in any form, you must be very patient with them. Reassure them that you love them and be kind, lighthearted Mr. Nice Guy. Whatever you do, don't hit them. Let your wife be the disciplinarian for a while longer. Give your tots a chance to vent their anger as you act cool as a cucumber. Eventually, they will have no other option but to love and "forgive" you if you show a little patience and restraint.

Your friends.

These folks will want to know everything that happened from their awesome war hero buddy. "How many people did you kill? How did you kill them?" No story will ever satisfy them, so always leave them wanting more. Whenever someone rudely asks me if I ever killed anyone in the war, I use the line from the film *True Lies*:

"Yeah, but they were all bad."

Then I change the subject. Less is more and it is no civilian's business what happened in combat.

Regaining your civilian bearing.

Slow and steady wins the race. That is a cliché, true, but it certainly is appropriate for what mentality to have after the parades are over and everyone may have seemed to forgotten what you did for our Nation. You may have frustration when driving. This is normal, as there are a lot of jerks on the road. But you can not throw them out of their cars or blow out their tires. This is frowned upon in modern society.

Try to avoid racist comments to Middle Easterners who now live in the USA. They are Americans, and left their harshly-ruled little countries to be part of our team. Can you blame them? Besides, most of them had to grow up in war, escaping with just the clothes on their backs. As I mentioned before, VA therapy is available to you at no charge, for as long as you need it.

Dealing with cops.

These chaps deserve a little respect in their own right, because their war is on the streets. Most of them are cool to vets, but there are more than a few officers who are bitter about missing the chance to kill terrorists. If you aren't breaking any laws and receive harassment anyway, keep your cool, and follow these three steps:

1) Be polite,
2) Get their badge number and
3) file a report against the city.

Media exposure.

When you feel up to it and if you are a ham like me, call your local newspaper and tell them you are back in town to give them an exclusive on your experience. Try not to bash the military; they get enough bad press here by lifelong spoiled college brats. Anything you know that is still classified should stay that way to avoid jeopardizing the troops that are still in battle. Everything else is free reign. It would be good of you to mention your spouse, family/friends and anyone nice enough to send you mail during your tour. Make sure to write down a few things you wish to mention well before the reporter stops by to chat. This way, if you draw a blank, simply look at your notes.

Bang out that degree. You earned it!

If you thought ahead and got hooked up with the Montgomery GI Bill during recruitment or basic training, use it. You have around 10 years to spend this money for college.

Here is how it works:

Visit a student VA representative, fill out the GI Bill forms and register at a community college for the first two years of your bachelor's degree. This way, the money for college stretches as far as possible. It is a big waste of money to get your undergraduate at a university, when the credits from a fully accredited community college are just as good and transferable.

Spend the last two years at a fancy college so you get their "name-brand" on your diploma. Try to do a full twelve credits every semester in order to receive the maximum amount of your monthly tuition checks. Then simply sift out all the useless knowledge of each class to make room for more. It's all just regurgitation of class notes and big business capitalism disguised as socialism. That's all college is in my opinion. The real stuff is learned on the job in most cases. Use school to get the next level career you want. I used unorthodox ways to get through certain areas of the required classes and didn't feel too bad after what I paid in text books. One of the best ways to get ahead in life is with a degree, so just dive right in. Whatever your situation, you will see students from all walks of life in attendance. But don't get too caught up in the learning or partying of the institution. The mission is that little piece of paper at the end of the road. Do what you need to pass each class and then make room in your brain for the next semester. Just having your Associate is not enough these days and a Masters or Doctorate is a lot of extra school with the probability of being overqualified. A bachelor's degree is enough to get recognized in the upper-class work force without the headache of giant student loans to pay back.

Enjoy the show.

You are the big man or woman on campus, and don't let any left-wing professor or anti-war tofu munching, stinky-armpit-braided protestor bum you out. Most these people have sheltered little lives in the fluff and fantasy land of academics to hide away in. They will spend their whole lives there, probably. You gave them the right to do this by putting your butt on the line. So just laugh off their pathetic little existence, to yourself.

Chapter 20: My words of farewell

Thank God for every breath. If you sat down and actually calculated the odds of coming back from the current wars after all the casualties the US has had since the 1990's, it's staggering that you are one of the lucky ones. You now have purpose and more to discover.

Do it!

Television is a huge waste of time. You and I have something else in common ... We both survived a war and we're still alive. Isn't that cool? All that matters is where you are right now, Amigo. Do not squander your precious time working a civilian job you hate because it pays well. Find your true passion in life and then pursue it. The money will come. If you are in a bad relationship or single and lonely, the right girl is going to appear in your life the moment you least expect it.

Pray for things to happen, and they will. God loves us very much, but He's nobody's genie. If your prayer is not answered right away, it's because He has something even better for you around the bend. PEACE OUT.

###

Meet the Author

Sebastian DiGiovanni has had an eclectic range of jobs:

Mortician's assistant, radio personality, theme park coordinator, talk show host, newspaper reporter, strip club disc jockey, and ten years in the military with back-to-back tours in the combat theater of Iraq and a hardship tour in Korea. His duties in the service were infantry, military intelligence, mortars and recon.
He is a licensed hypnotist; also, licensed to "marry you and bury you." Translation: he can perform weddings and preside over funeral services.

He lives in Arizona with his beautiful and talented partner Hydie and their three amazing kiddos.

Most of Sebastian's writing stems from the odd career paths he's chosen.

Works in progress include his debut fiction work, *Anodes* (pronounced A-nod-es) *Diner*, a collection of short stories reminiscent of *Twilight Zone* and *Outer Limits*; also, a novel, *The Locket Diaries*, a story of love and war set during the war in Iraq.

Here is one of Sebastian's short stories from *Anodes Diner*: (Watch for news of its publication at http://www.marjimbooks.com; also, learn about other good books published by UCS PRESS)

Table #4: Bill's Companion

Always the gentleman, Bill carried his new bride over the threshold of the diner. The hostess saw he had his hands full and led him and his wife to a booth, where he carefully set her down. After a little pleasant patter with the waitress, she left. Water was all he wanted, not for the missus, just a full glass after a long drive.

Bill gazed into his soul mate's eyes from across the booth and his heart jumped as it always did, for he was sitting across from Jenny. He thought back to the studio where they first met…

"Good evening and thank you for joining us at Vision Core News. I'm your host, Jenny Kline, with headlines from June 18th, 2012."

How those eyes sparkled, like marbles in the sun. Her skin was unblemished and lips as rich as merlot. She could announce the earth being knocked out of orbit and it would still bring appeal to one admirer. A lonely, middle-aged fellow known only as Bill reached climax as she paused for a commercial break; and gave him enough time for a cigarette before returning with more news of the world.

The studio wasn't far from his apartment in Burbank. He had been to the studio on numerous occasions, with no luck of spotting her. On Bill's next visit, he would wait. As long as it took to spot her, he would wait. After all, this was his only companion at 6 o'clock and any other hour day or night. He owed her a visit to tell this woman how much she meant to him. Time was always a factor in a state that could be swallowed up by the ocean due to 'quakes. Probably from all the weight of data National Geographic people collected over the years. At least that was Bill's assumption of his fellow Californians. He kept these periodicals and many others stacked in his cluttered abode and figured everyone else did the same.

The tour of this impressive facility was always a bore to him. Bill didn't care about an old set built for seven stranded castaways, or James Bond's Fort Knox. However, the full scale model of a killer shark from some summer blockbuster in the 70's did hold his interest long enough for Bill to punch it in the decaying foam head while the tour guide wasn't looking. This antique monstrosity terrorizing the beaches of Amity was the reason Bill could not even so much as swim in a hotel pool without paranoia since the movie's release. He felt a bit of closure with each swing, on every tour.

Bill's exit was approaching. When he saw the familiar VisionCore sign, this was Bill's cue to discretely break off from the pack. He paid for the tour and this time, nothing would stop him from meeting his Jenny. Even a mere glimpse of her this time would satisfy Bill's obsessive behavior.

Bill sat down on his usual couch in the lobby. He was the type that never got a second glance throughout his entire monotonous life. For this occasion, it served him well to be hum-drum. Security and the receptionist had not so much as given Bill a second glance, as usual. Perfect. VisionCore was always buzzing with little worker ants. Production assistants, stage-hands and grips are always so important in their own minds. Little did these folks realize, their jobs were one coffee stain on a director's crotch from being terminated.

She wouldn't be hard to spot and Bill had all the time in the world. His walking about all day made the couch in the reception room extra cozy this time. The tired man let his heavy eyes drop for the evening.

Waking with a start and glancing around, Bill had a small panic attack when he realized he wasn't in his own bed. Glancing around nervously, the details of his daytrip came back to him with a sense of relief.

The VisionCore sign was the only light source illuminating the lobby.

How long had he been asleep?

Darn it, he missed another chance to meet his dream girl. Or was he still dreaming?

"Awe, piss on it!" Bill muttered to himself. If he were still asleep, his back wouldn't be so sore. His only day off from his work as an electrician this week and he spent it without so much of a glimpse of her. Another thought crept up on Bill like a wet fart. His time going unnoticed through life had finally paid off. Bill could have a look around where Jenny worked without anyone telling him an area is off limits. Not even one security guard in the area. Bill had his run of the place.

The door to the soundstage was locked. Hardly a problem, as Bill dabbled in locksmithing as a hobby. He grinned as he felt the tumblers in the lock gently give way and release the deadbolt. Bill opened the door. The set was dark. He fumbled for a light switch. Knowing that if caught, Bill could pretend to be mentally impaired, a talent that once got him out of paying for his ticket on the train when a conductor came by checking.

Bill froze.

There was someone sitting in the news anchor's chair across the room. Could it be? Perhaps Jenny fell asleep in the same manner as Bill, getting locked in with him.

"Wow!" he thought. "We are soul mates after all."

No longer thinking of consequence he flicked the switch to the stage lights. It was Jenny alright... But there was more. Bill took a step back, having trouble breathing. The harsh lights revealed what television viewers at home were never going to be privileged enough to see. There was a reason camera level never dropped below Jenny's waist. She didn't have one. Or anything below her waist that resembled human. Just a two-foot steel pole bolted to the floor, wrapped in wires and pistons.
Bill wondered no more about why the producers had chosen not to show off this beautiful woman's legs. They hadn't designed her with any.

Jenny had the shape of a seamstress' mannequin, if you added a head and arms. Her desk with the familiar station's logo had been moved and a technician's tool box was in its place next to her base. As he walked over to this modern marvel of science, Bill tried to grasp the fact that he had been in love with a machine all along. No wonder her skin was unblemished, it was made of silicone. The merlot of her lips had been carefully painted on and the reason Jenny's eyes sparkled like marbles is because they probably were.

Still in shock and equally fascinated at the same time, Bill followed the wires to a console. The control panel was intricate, but everything of importance was clearly marked. He turned a dial marked *INTRO*. Jenny's facial expressions sprang to life.

"Good evening and thank you for joining VisionCore News. I'm your host, Jenny Cline, for…"

Click.

Bill assumed the current date and time had to be programmed in for each broadcast. He located the *SETTINGS* button and pressed it. Jenny announced it to be Monday. She reminded him of an alarm clock radio when he continued to tap the button. "Tuesday, Wednesday, Thursday…" she announced, simulating a breath in between. That was a very lifelike touch the engineers added, caused by compressed air through a tube running up Jenny's skirt and into a set of artificial lungs.

Looking at the control board, Bill noticed the news anchor could be placed in *CONVERSATION* mode. He figured this purpose served for interviewing celebrities. Yet another reason for the large desk that usually covered her lower half.

"To keep the network's dirty little secret," Bill mumbled.

The prop designer even added small curtains to hide Jenny's lower half from any prying eyes on the set. This was like some bizarre twist to *The Wizard of OZ. Pay no attention to that man behind the curtain. I am the Great and Powerful Jenny Cline, with today's national weather report.*

Bill didn't care anymore. He loved her, machine or not. He flipped the switch to *CONVERSATION* and waited. Jenny looked over at the booth where Bill was standing. "Hello… I'm Jenny Cline. Who might you be?"

Bill looked at her with more surprise, but answered, "Hi Jenny, I'm Bill. It's really a pleasure to make your acquaintance." Her number one fan still couldn't get over what a life-like replicant she was.

"The pleasure's mine, Bill. Are you my new technician?" she asked in the friendly tone Bill was so used to when all alone in his apartment. He pondered her question for a moment, as he looked at the tools lying beside her.

"Would you like me to be?"

***

To Whom It May Concern,

This doubles as letter of resignation and an apology for stealing.

Sincerely, Bill and Jenny